MEASURES

James M. Young

Robert L. Barth
14 Lucas Street
Florence, Kentucky 41042

For Meg and Kate

"Thus love I still"

"For My Parents," "For Both of Us," "A Song for the Commonwealth Games," and "Martial 12.10" first appeared in THE COMPASS: A PROVINCIAL REVIEW. Earlier versions of "On the Death of a Musician" appeared in THE COMPASS and THE GRAMERCY REVIEW. "To Accompany a Collection of Recipes" appeared in a pamphlet compiled by Meg Young.

ISBN: 0-941150-11-9

FOR MY PARENTS

In short, advice
Will not suffice
Though it be brief
And filled with wit,
For him whose face
Lacks learning's grace,
And life, the grief
That comes with it.

FOR BOTH OF US

This day, four years have passed, and in that time,
As minutes merge into continuum,
So have our separate thoughts and actions merged
Into a flow of mutual circumstance.
We are not of one mind, and will not be;
But still, distinctions fade as do the hours
Into the past, and presently this too,
Though mine alone, will seem conceived of you.

ST. AUGUSTINE

Ships from the north, riding the wine-dark sea,
Remind me of Imperial Rome in flames.
This is our end, but not the end to be;
Infinity plays no such paltry games.
  I am too old, we are too few to fight;
  Armor yourselves, my children, with the Light.

A WESTERN PROMONTORY

Out from the Cornish coast a rock extends,
Held to the mainland by a single strand;
Massive and sheer, it overrides the sea.
Along the lower cliffs the gulls can land

Above the breakers' spray, and there can rest,
Or shelter on the south side from the storms
That roll in from the Irish Sea to lash
The upper flats. Here nothing grows but forms

Of lichen and, in crevices, dry grass.
In the gray rock below the bare plateau,
The outline of their hovels can be traced
Who lived there thirteen centuries ago,

Driven, by a desire to sever will
From flesh, to settle on the seaward face,
That spirit might forget the single strand
In contemplation of a changeless place.

The ocean swells, subsides. The constant roar
Is carried with the wind along the coast
And, inland, fades; at the edge, even in sleep
The incessant rise and fall would linger most,

While they who lived in darkness dreamed of light, —
Convinced they would not pass that way again, —
Until distracted by the darkening line
Where land and ocean fade away in rain.

SOMETIME NEAR SUMMER

One day I biked out farther than before,
To where a black house, heading for the ground,
Sagged in the spear grass. Poplar and low brush
Bordered the fence; but leaning in the door,
The goldenrod seemed waiting to know more,
And earth and log were ready to compound.
Inside, the warm air hardly held the rush
Of scurrying below the rotted floor.

Up all four walls old newspapers were lined
In ragged, overlapping strips of gray.
I read the lower bits, amazed to find
That in the crumbling commerce I could trace
The dispersed pleasures of the commonplace,
Deep in their proper business of decay.

ON THE DEATH OF A MUSICIAN

This is the gift: to reach the most remote;
To seek through every age the vibrant note,
Fix on it, study, make it wholly yours,
From flaking codices draw living scores;
To set and play them all, motet and mass;
To live again the minds that none surpass.

And was that not enough? Did such skill teach
But an impatient rage to overreach
Their mastery? Massive Leonin, Dufay,
Susato, Dowland, and Machaut—were they
All nothing to you; or, lost to your sight,
Forms that withdrew as you fell into night?—

The night that nothing shows. In that last daze,
What demon held you with his empty gaze?
Hearing you now, this I can understand:
You died at thirty-three by your own hand;
Composed from what you knew a final choice—
The long melisma of the breathless voice.

A SONG FOR THE COMMONWEALTH GAMES

Edmonton, 1978

The infield green is bare;
The seats are full. Flags crack
In the quickening air.
The sun expands the track.

Within the stadium
All seeing is defined.
Beneath the immense, low hum,
In lanes precisely lined,

Picked runners will compete;
Where discuses are spun
Each grass blade seems discrete.
The eye sees all in sun.

From the high, plunging stands
It bends out street by street
To where the plain expands,
Blurred in the rising heat;

There, edged by pure surmise,
Miles melt away from sight;
Dark summer fallow lies
Lost in the summer light.

Here the ancient games unfold:
Here vendors hawk their bait,
Here athletes strain for gold;
While others contemplate—

But with no heavier duty:
In this confined release
To watch hard, sinewy beauty
Flux with elusive ease;

Beguiled and unreproved,
To be in the body's hone
Entranced, in motion moved
For no sake but its own.

EPIGRAMS

1. Martial 12.10

Grasp, buried in his gold, beggars his son.
Fate gives too much to many, enough to none.

2. Martial 5.73

You ask my verse, so here. This evens scores:
I had kept mine in hopes you would keep yours.

3. *Carpe* knocks it back to keep the past at bay,
To put tomorrow off, and kill today.

4. Martial 1.28: On Acerra

Fetid from last night's wine? You're too far gone
Yourself: you know he always drinks till dawn.

5. Martial 6.19: To Postumus, the Lawyer

It's neither rape nor murder nor a draught
Of poison that concerns us; I complain
Of three she-goats lost by my neighbor's craft.
The judge wants just the plain facts of my case.
But you, good sir, boom forth in tragic strain
Of Cannae; gesture; wander from your notes
To cry rage at the perjured Punic race;
Damn Sulla, Marius and Mucius praise.
Could you work in, somewhere, my three she-goats?

6. Martial 3.9

Cinna, you say, writes squibs at me? Not quite.
No one reads Cinna: Cinna does not write.

7. I am with *Suave* against the common hoard.
When he gives public readings in that tone
Of cultivated boredom so adored
By all, I empathize with him alone.

8. Martial 6.41

Cinna is hoarse, yet will not cease:
He neither speaks nor holds his peace.

9. Martial 1.64

You're pretty, young, and rich. Who would deny it?
But when you say you are, girl, I deny it.

10. With the Concordance to Ovid

Forgive us, Ovid, who are so perverse
As bend our backs with this to read your verse.
How could a book be on your being centred
Where naked thighs are entries and not entered?

11. To Accompany a Collection of Recipes

Now that December has arrived, we send
No neckties, stationery, cash to spend;
No candles, watches, or dried fruits—we doubt
You need new art or socks; and Guinness Stout
You'll buy yourself—no chocolates, nuts, liqueurs,
No sets of handkerchiefs marked "his" and "hers";
No coffee-table books, no Christmas gnomes,
No bargain shirts, and not one book of poems.
We send no gift that asks a gift again.
Instead, here is a thing, homespun and plain:
Some recipes collected over time.
These, if you'll add some Californian wine,
A few close friends, familiar talk, and free
Occasion to enjoy, we guarantee.
For though the whole is ours, from you we took
Each recipe here entered in this book,
Which now we give as only your just due:
If gifts reveal the giver, this shows you.

ON A PHOTOGRAPH, C. 1940, SEEN SOME TIME AGO

In memory, M.A.R.Y.

Older than I, the yellowed photo stands
As what you were before I knew you. Bent
To the table, you open up a patient,
Anesthetized and calm beneath your hands.

A watchful nurse, a tent of bulky linen,
And, masked, your half-seen face: the details blur.
Perhaps the instruments glint silver-white;
Perhaps I think them there merely to fill in

The grainy space below your hands. But they,
Clearly, are there. Each fingertip is light.
Steady in action and in memory,
They curve and pause, articulated, sure.

Now even steadier, at their last stay,
They move their surest in a long decay
That every concentrated skill you knew
Looked to delay in others for a time.

Full in yourself, you fill the picture—you
And a clock that shows twelve-twenty on the wall.
Lunch or past midnight? In the neutral glare
You work, as if such spareness were your all.

One hundred signed and numbered copies
of this edition have been published.
Twenty-five are lettered A to AA by
the author and are not for sale.
Seventy-five are signed and numbered by
the author for sale.
This copy is number ___10___.

James M. Young